ALL ABOUT ANIMALS

DOGS

CHELSEA HOUSE
PUBLISHERS
A Haights Cross Communications Company ®
Philadelphia

Library of Congress Cataloging-in-Publication Data
ISBN: 0-7910-8686-0

The series *All About Animals* was created and produced by:

McRae Books Srl
Borgo S. Croce, 8, 50122, Florence (Italy)
info@mcraebooks.com
e-mail: mcrae@tin.it
www.mcraebooks.com

Publishers: Anne McRae and Marco Nardi
Text: Chris Hawkes
Illustrations: Alessandro Baldanzi, Fiammetta Dogi,
Paula Holguín, Lucia Mattioli, Leonardo Meschini,
Antonella Pastorelli, Studio Stalio (Alessandro Cantucci,
Fabiano Fabbrucci, Margherita Salvadori)
Graphic Design: Marco Nardi, Yotto Furuya
Layouts: Rebecca Milner, Laura Ottina
Picture Research: Helen Farrell, Chris Hawkes, Claire Moore
Editing: Claire Moore
Colour separations: Litocolor, Florence (Italy)
Cutouts: Alman Graphic Design

Printed and bound in China

Acknowledgments
All efforts have been
made to obtain and provide
compensation for the copyright
to the photos and artworks in
this book in accordance with
legal provisions. Persons who may
nevertheless still have claims are
requested to contact the publishers.

The publishers would like to thank the following sources for
their kind permission to reproduce the photos in this book:
t = top; tl = top left; tr = top right; tc = top center; c = center; cl = center left; cr = center right;
b = bottom; bl = bottom left; br = bottom right; bc = bottom center

Art Wolfe: 16bl; with thanks to Carolyn Scott: 26c;
Contrasto/Corbis: 17cr, 21, 24–25b, 25cr, 26–27b, 27tr, 29, 32tr;
©Disney Enterprises: 32bl, 32cl; Farabola Foto (Bridgeman Art
Library): 23br, 26tl; Marco Nardi: 19c, 35tl, 36br, 36cr, 37br, 37cr;
National Portrait Gallery, London: 33cr; Panda Photo, Rome: J.
Foott/Panda 17b; The Image Works: ©Eastcott-Mornatiuk/The
Image Works 9tl, ©The British Library/Topham-HIP/The Image
Works 11tr, ©Nancy Richmond/The Image Works 19br, ©Bob
Daemmrich/The Image Works 19tr, 37tr, ©Syracuse
Newspapers/David Lassman/The Image Works 24tr, ©Jenny
Hager/The Image Works 25tr, ©Thomas Wanstall/The Image
Works 25cl, ©Syracuse Newspapers/Randi Anglin/The Image
Works 27tc, ©Topham/The Image Works 28bl; "SCOOBY-DOO
TM & © Hanna-Barbera Productions. All rights reserved." -
(Omega/Cinetext): 32

p. 8: extract from 'The Cat that Walked by Himself', a short story
by Rudyard Kipling (1902); p. 14 extract from 'The Unbearable
Lightness of Being' by Milan Kundera (1984); p. 16: extract from
'A Celebration of Dogs' by Roger Caras (1984); p. 18: quotation
by Plato (427–347 B.C.); p. 22: extract from 'The Talisman', ch. 14,
by Sir Walter Scott (1825); p. 24: extract from 'Idle Thoughts of
an Idle Fellow: On Cats and Dogs' by Jerome K. Jerome (1886); p.
27: quotation by John Steinbeck (1902–1968); p. 31: extract from
'The Devil's Dictionary' by Ambrose Bierce (1911); p. 33: extract
from 'Snoopy' by Charles M. Schulz.

ALL ABOUT ANIMALS
DOGS

Chris Hawkes

List of Contents

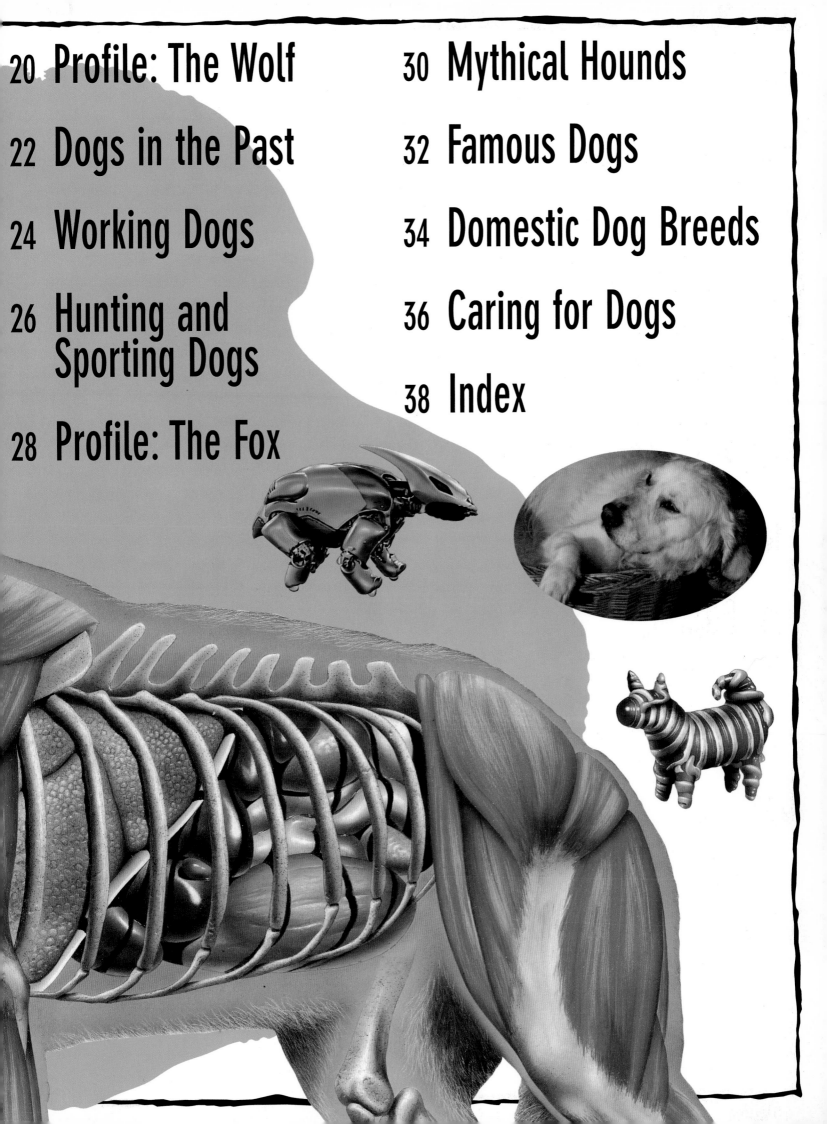

What Makes a Dog a Dog?

Dogs are the oldest of the carnivores (from the order *Carnivora*) and were the first animals to be domesticated by humans. Fossil evidence leads scientists to believe that canines have been around for about 40 million years. All dogs, from the giant Saint Bernard to the tiny Chihuahua, are thought to be descended from the Gray wolf (*Canis lupus*).

When wolves bark at the moon, they are trying to call the pack together.

➲ Barking

All dogs, whether they are domestic or wild, bark as a means of communication. For some reason, domestic dogs can bark louder than wild dogs. The Basenji, an African wolf dog, is the only dog that cannot bark. It was bred hundreds of years ago as a silent hunting dog.

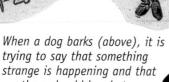

When a dog barks (above), it is trying to say that something strange is happening and that others should be alert.

All wolves, including this Gray wolf (left), live and hunt together in groups called packs.

✆ Pack instinct

Dogs are social animals and have a natural instinct to be a member of a group. This helps explain why the dog has become the perfect pet, as the human clan provides a dog with what it needs most — security and attention. It also explains why, if a dog loses its master, it has to find a new one.

↻ Dogs and cats

Despite popular myth, the chances are that dogs and cats living in the same household will get on well. Dogs are social animals by nature and love companionship — be it other dogs, cats, or humans.

This stained glass window (right) from the Middle Ages shows a pet dog asleep on its owners' bed.

Most dogs are happy to be petted and fussed over.

◑ Faithfulness and loyalty

Dogs instinctively feel the need to live their lives within a social hierarchy, either as one of the subordinate members of the group or as its leader. For the dominant owner, a dog will give endless love and loyalty. The ease in which a dog will become subordinate depends very much on the breed that is belongs to.

This stamp (above) was issued by the British Post Office to commemorate the paintings of the British artist George Stubbs (1724–1806).

◑ Agility

Most dogs have evolved into naturally athletic animals, largely through the need to catch prey and to avoid predators in the wild. Domestic dogs — especially the larger breeds — also need plenty of exercise. Most can be trained to perform acrobatic feats such as leaping through tyres.

◑ Dogs and people

Fossil evidence suggests that members of the dog family and people have been living in close contact with each other for about half a million years. How and when wild wolves became domestic dogs, however, is unclear. The most likely explanation is that wolves started to gather around human settlements to scavenge for food and eventually became tame.

◒ Big eaters

Dogs are primarily meat eaters, but they require a balanced diet if they are to lead healthy lives. The amount of food that a dog eats depends on its size — a wolf in the wild can eat up to 9 pounds (20 kg) in one meal.

◒ Hunters

All wild dogs, including wolves (left), are natural hunters and will actively track, chase, and kill prey. Some pet dogs still have this hunting instinct and have been known to attack livestock. As well as hunting, most wild dogs are also scavengers, feeding on carrion (dead meat), and many also eat berries, fruit, and grass. Wild dogs often store excess food by burying it, in the same way that pet dogs bury bones.

The Dog Family

Although today's domestic dogs are vastly different from their wild ancestors, they all belong to the group *Carnivora*, an order that dates to 55 million years ago. Ever since people first domesticated the dog, their attempts at cross-breeding have dramatically altered the natural course of the dog's evolution.

Hesperocyon (left) was the first dog-like creature to walk the earth and lived in North America.

🎧 Migration

The first members of the modern canine family appeared in North America around ten million years ago. Some migrated into South America where they evolved into foxes. Others crossed the Bering Straits (which was then land) into Asia and spread into Europe. After the first dogs had been domesticated, they accompanied humans as they moved around the world.

➲ Early ancestors

The first true dog-like creatures were called Hesperocyon. They evolved in North America around 40 million years ago and became extinct about 15 million years ago. Hesperocyon evolved into two types of dog on either side of the Atlantic Ocean: Cynodictis in Europe (which became extinct 35 million years ago) and Pseudocynodictis in North America, which evolved into *Canis lupus*, or the Gray wolf.

The lower jaw of the Cynodictis (above) bears a close resemblance to that of the modern dog.

He is your friend, your partner, your defender, your dog. You are his life, his love, his leader. He will be yours, faithful and true, to the last beat of his heart. You owe it to him to be worthy of such devotion.

Unknown

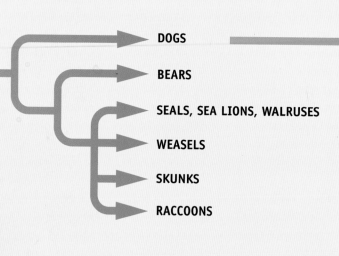

An artist's impression of Tomarctus (left), an ancestor of the modern dog that lived during the late Miocene period.

FAMILY TREE OF CANINES

Dog family tree

A lack of archaeological evidence means that we cannot be certain about the exact lineage of the modern dog. This diagram shows the dog's most likely evolution, although not all scientists would agree with it.

- DOGS
- BEARS
- SEALS, SEA LIONS, WALRUSES
- WEASELS
- SKUNKS
- RACCOONS

- CATS
- CIVETS
- MONGOOSES
- HYENAS

40 million years ago

30 million years ago

HESPEROCYON

BOROPHAGINES

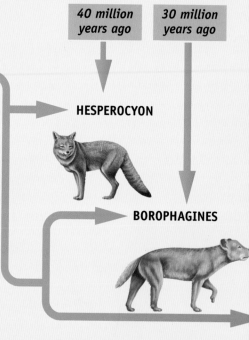

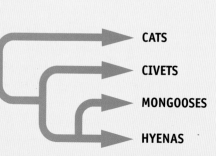

Evolution of the Dog Family

⊆ The Gray wolf — the dog father

Despite a lack of fossil evidence, scientists generally believe that the Gray wolf is the ancestor to all modern breeds of dog. It is thought that these animals started to hang around the settlements of Stone Age people, scavenging for food and so began the domestication process. Humans then began to breed these wolves selectively, keeping the characteristics that they desired.

A scene from The Jungle Book by Rudyard Kipling shows Mowgli and wolves.

Scientists believe that the Gray wolf (above) is the most likely forefather to the modern dog.

⊃ Dog species

Selected cross-breeding of dogs began at various places around the world at different times. By the Bronze Age, there were five definite types of dog. Most of the breeds of dog that we see today have been with us since the turn of the 20th century, with only a handful dating back further. Today, the Kennel Club of the United Kingdom recognizes 192 breeds of dog.

Early drawings, such as this one found in Libya (left), show that early people and dogs were hunting companions.

Above: By the Bronze Age there were five categories of dog: (1) Canis Familiaris Palustris; (2) Canis Familiaris Intermedius; (3) Canis Familiaris Metris Optimae; (4) Canis Familiaris Lenieri; and (5) Canis Familiaris Inostranzevi.

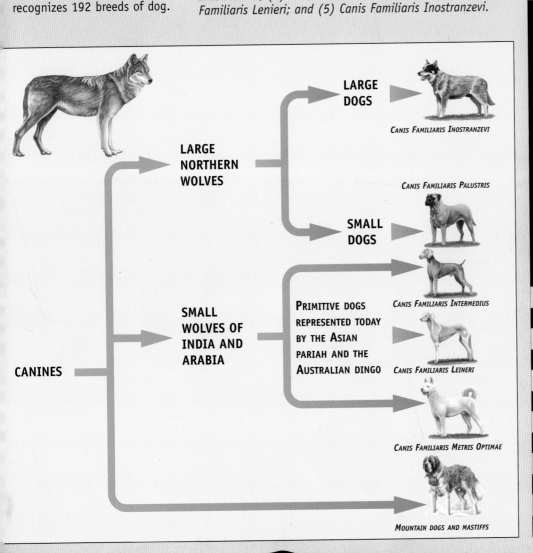

CANINES → LARGE NORTHERN WOLVES → LARGE DOGS → *Canis Familiaris Inostranzevi*

LARGE NORTHERN WOLVES → SMALL DOGS → *Canis Familiaris Palustris*

SMALL WOLVES OF INDIA AND ARABIA → PRIMITIVE DOGS REPRESENTED TODAY BY THE ASIAN PARIAH AND THE AUSTRALIAN DINGO → *Canis Familiaris Intermedius*

Canis Familiaris Leineri

Canis Familiaris Metris Optimae

Mountain dogs and mastiffs

Shared characteristics

All dogs have a pair of very sharp teeth called carnassials in both their upper and lower jaws. They help the dog slice through muscle and skin and are common to all carnivores.

CANIS FAMILIARIS INOSTRANZEVI
Akita, Alaskan Malamute, Elkhound, Samoyed, Siberian Husky, Keeshond

CANIS FAMILIARIS PALUSTRIS
Chow Chow, Eurasier, Finnish Spitz, Pomeranian, Bull Terrier

CANIS FAMILIARIS INTERMEDIUS
Basset Hound, Beagle, Bloodhound, German Short-haired Pointer, Braque Français, Poodle, Cocker Spaniel

CANIS FAMILIARIS LEINERI
Pharaoh Hound, Deerhound, Greyhound, Irish Wolfhound, Afghan Hound, Saluki

CANIS FAMILIARIS METRIS OPTIMAE
German Shepherd, Berger Picard, Collie, Beauceron, Briard, Old English Sheepdog

MOUNTAIN DOGS AND MASTIFFS
Pyrenean Mountain Dog, Dogue de Bordeaux, Komondor, Leonberger, Mastiff

A World of Dogs

Dogs have evolved into both a highly adaptable and extremely diverse group. Wild dogs, in all shapes and sizes, can be found in many different parts of the world, in environments as diverse as the freezing wastelands of the Arctic tundra, the tropical rain forests of South America, and the searing heat of the Sahara Desert.

⊂ Arctic fox

Found in the far north, mainly in the Arctic Circle, the Arctic fox has a thick coat and hair on the pads of its feet to protect it against the freezing temperatures. It feeds on voles, lemmings, hares, birds, and birds' eggs and stores food in the summer for the winter months.

↻ Bush dog

This little-known dog lives in the forests and marshlands of Central and South America. With its sturdy body, short tail, and small ears, it does not look much like other dogs. Bush dogs hunt together in packs, often driving their prey into the water, where they can kill it more easily.

∩ Gray wolf

The Gray or Timber wolf inhabits northern North America and Asia. It lives in packs of around 20 and is thought to have one mate for life. Gray wolves are the largest members of the dog family, the male sometimes measuring 6 feet, 6 inches (2 m) in length and weighing up to 110 pounds (55 kg).

∩ South American zorro

There are four species of the fox-like zorro spread through South America. Their habitat ranges from rain forests to grassy plains and low mountain areas. Their diet varies from small mammals to fruit. They have been hunted extensively for their fur. One million zorros were killed in Argentina alone in 1978.

⊃ Coyote

The coyote takes its name from the Aztec word *coyotl*. It lives in North and Central America, from Alaska to Costa Rica. Known for its yapping at night, it has adapted well to human settlements, increasing its range in recent times. It feeds on rodents, hares, carrion, and vegetable matter.

∩ Maned wolf

The maned wolf, now quite rare, lives in the isolated grasslands of central South America. It has long, reddish fur and very long, dark-colored legs. A loner, it hunts at night, feeding on small animals, insects, and some plants.

⮂ African wild dog

Also known as the Cape hunting dog or the Hyena dog, it lives in southern Africa. Its blotchy black, yellow, and white fur helps it blend into its grassland home. African hunting dogs live and hunt in packs of up to 60 individuals.

⮂ Dingo

The dingo is an Australian wild dog. It was brought to the continent from Asia about 8,000 years ago by aborigines. Dingoes hunt alone or in small groups, feeding on kangaroos, rabbits, and stock. They can be tamed and sometimes make affectionate pets.

⮂ Bat-eared fox

The Bat-eared fox is the only member of the canine family to have abandoned meat in favor of insects. Found in southern and eastern Africa, they are nocturnal animals that tend to live in pairs. Although hunted extensively for their fur, they are not an endangered species.

↻ Raccoon dog

Found in the northern areas of Asia, Raccoon dogs live in forests close to a source of fresh water. They are excellent swimmers and their diet is made up mainly of fish. Primarily nocturnal, they live alone or in groups of up to six members. They have a dark patch on their face, like raccoons, after whom they are named.

↻ Dhole

Native to western Asia, India, Indochina, and Java, it is rare to find a dhole in the wild. It is a forest-dweller that tends to live near to a source of fresh water. Its dark tail stands out sharply against its sandy, russet fur. The dhole's diet consists of deer, birds, and lizards.

↻ Fennec fox

The Fennec fox lives in the deserts of North Africa and the Arabian peninsula. It has a small body with large ears and a long, black-tipped tail. Fennec foxes are mainly nocturnal hunters. They stay in their cool burrows during the day.

⌒ Jackal

There are three species of jackal — the Asiatic, the Black-backed, and the Side-striped. The Asiatic jackal lives in Eurasia and North Africa. The other two live in Africa. Jackals hunt at night, feeding on carrion or small animals. They have a strong offensive odor.

⮂ Red fox

The most widespread of all wild dogs, the Red fox is the largest member of the fox family and can adapt to a number of different environments, from deserts to city centers. Despite their name, Red foxes can come in a variety of different colors, from red to pure white to black.

⊃ Smell

Some scientists believe that a dog's sense of smell can be anything up to one million times more powerful than that of humans.

> **Dogs are our link to paradise. They don't know evil or jealousy or discontent. To sit with a dog on a hillside on a glorious afternoon is to be back in Eden, where doing nothing was not boring — it was peace.**
>
> *Milan Kundera (1929–), author*

Truffle dogs have such an acute sense of smell that they can detect a truffle that is 3 feet, 3 inches (1 m) underground.

⊃ Skeleton

A dog has approximately 320 bones that provide the frame for its body and help protect its major organs. There are four types of bone: long bones (in the legs); flat bones (in the shoulder); irregular bones (vertebrae); and short bones (such as in the dog's paw).

⊂ Sight

Dogs are thought to be color blind. They see objects first by their movement, then by their brightness, and finally by their shape. Dogs have a light-reflecting layer behind their retina which acts like a mirror and helps them see in twilight. It also makes their eyes glow in the dark.

⊃ Teeth

A fully-grown dog has 42 teeth made up of 12 incisors, four canines, 16 premolars, and 10 molars. A young dog will only have 28 milk teeth, but will develop a full set by six or seven months. You can tell the age of a dog by looking closely at its teeth.

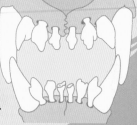

On average, a dog's mouth can exert up to 150–200 pounds (68–91 kg) of pressure per square inch. Some dogs can apply up to 450 lb (204 kg) of pressure.

Different dogs carry their ears in different ways according to the characteristics of their breed. A German Shepherd dog's ears point upward, whereas those of a poodle droop and hang flat against the side of its face.

↻ Digestion

The dog's digestive system comprises all the organs required for the dog to ingest food and to transform it into the energy it needs to move and stay warm. Unlike humans, the dog does not have an appendix.

Muscles are attached to the bones by tendons and allow the dog to move. A dog also has cardiac muscles (the heart) and smooth muscles (in the digestive system).

∩ Circulation and respiration

Blood, which provides the dog with nutrients and oxygen, is pumped through a network of veins and arteries by the heart. Two lungs provide the dog with oxygen and remove poisonous carbon dioxide from its bloodstream.

The Dog's Body

The striking thing about dogs is that they can be so different — in size, shape, color, and temperament — yet they are all dogs. The tiny Chihuahua (the smallest dog in the world), weighs just 2–7 pounds (1–3.5 kg), while the Great Dane (one of the largest dogs) weighs about 120 pounds (55 kg). Wild dogs also differ greatly in size and color. Even so, all dogs have powerful muscles, tough tendons, and a sensitive nervous system that allows them to move quickly.

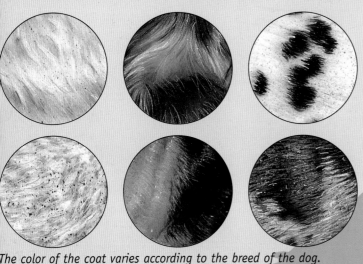

The color of the coat varies according to the breed of the dog.

The coat

A dog's coat helps the animal maintain a constant body temperature. It is usually at its thickest, and its longest, during the winter. Dog owners should never clip a dog's coat before the start of winter. They should also always dry a dog's coat if it is wet, because it will not work as efficiently in this state.

Sickle tail

The tail

Different breeds of dog carry their tails in different ways. Pointers hold their tails straight and horizontal, whereas the tails of Chow Chows are usually curled. A dog's tail helps the animal maintain balance, which is important if the animal is running, jumping or swimming. The way that a dog carries its tail is also a good indicator of its mood and general health. Some dogs, like the Old English Sheepdog, are born without a tail.

Sword tail

Thermo regulation

The normal body temperature of a dog is 101.3°F (38.5°C). It has many ways of maintaining its body temperature. A dog's coat will trap warm air in the body to maintain heat. If a dog gets too cold it will start to shiver, an action that also produces heat. Since a dog has few sweat glands, water is evaporated through the mouth when it gets too hot, which is why a dog pants in the heat or after exercise.

Walking

A dog is a digitgrade animal, which means that it walks on its toes. It has four ways of moving, or gaits. The slowest gait is the walk when a dog moves its right front paw, followed by its left hind paw, then its left front paw and so on. The next speed, the trot, is the one used by most hunting dogs. The amble is the third fastest and is a dog's most natural gait. The gallop, with both the front and hind legs moving at the same time, is the fastest gait.

Curled tail

The number of vertebrae in a dog's tail varies according to the different breeds. In competitions, the judge will look closely at how the dog carries its tail. If it is carried properly, it means that the dog's pelvic muscles are in good order.

Plume tail

The shape of the tail and the way that it curves varies greatly from breed to breed. Here are a few examples (right).

Docked tail

A dog's stomach can hold between 1 and 9 quarts (1–9 liters), depending on the size of the dog. The dog's stomach is lined with glands that secrete enzymes that help break down the food.

Dogs have five toes on their front paws and four toes on their hind paws. The paws can either be normal, oval, round, or long. Some dogs have webbed feet to help them swim.

Gay tail

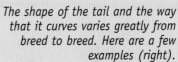

A Dog's Life

A dog's life in the wild is geared toward survival. Dogs are social animals that live together in packs. This makes them more successful hunters and helps them to protect themselves against predators. The packs are usually organized according to strict hierarchies.

Dogs in the wild fight to try to establish dominance over each other.

⮂ The leader of the pack

A pack of dogs needs both a structure and a leader to survive. The strongest member of the group will invariably become its leader. In the wild, the leader has many responsibilities toward the other members of the pack, such as defending their territory and finding food. Sometimes there will be a struggle for the leadership and fights may occur, particularly if food is short or if a female member of the pack is in heat (ready to reproduce).

These wolves (left) from northern Canada are fighting over the carcass of a dead deer.

⮂ Social hierarchy

All dogs, whether they are wild or domesticated, will try to establish a hierarchy. Domestic dogs will try to gain dominance over other members of the household from an early age, whether they are other animals or humans.

↻ Hunting as a pack

Wild dogs eat almost anything. The wolf's favorite food is moose, deer, or caribou. To kill an animal that can be up to ten times their size, though, requires teamwork. Wolves will track a herd, pick out the young, old, or injured and trail it for hours until it becomes exhausted. Then they close in for the kill.

> Dogs have given us their absolute all. We are the center of their universe. We are the focus of their love and faith and trust. They serve us in return for scraps. It is without a doubt the best deal man has ever made.
>
> *Roger Caras (1928–2001), author*

☣ Body language and communication

A dog's communication system includes scent markings and body language (which includes moving the body hair, eyes, and tail), facial gestures, and eye contact.

A subordinate wolf will assume a less erect position, drop both its ears and tail, and lick the nose of the dominant wolf (above).

The howling wolf (left) has become an animal to be feared by humans and has inspired many myths and legends. When a wolf howls, though, it is only communicating with other members of the pack.

☾ Vocal communication

Dogs also communicate with each other vocally, using whimpers, whines, barks, growls, and howls. A howl may just be for fun or used as a greeting. It may also be a means of locating other members of the pack or used as a warning to say that a trespasser has entered their territory.

☣ The sense of territory

All dogs, whether they are domestic or wild, claim a piece of territory as their own by secreting a scent that tells other dogs to keep out. Dogs will then instinctively defend this territory against intrusion from other dogs or humans.

⊃ Stray dogs

Sadly, some pet dogs are abandoned by their owners and are forced to return to the wild. These are known as stray dogs. They quickly form groups (right) and usually live near human settlements to find food.

This wolf (above) has rolled over onto its back as a sign of submission to the dominant wolf.

This wolf (below) is scavenging carrion while the stray dogs in the photograph (above right) are feeding on refuse.

Puppies

All puppies are born in a fluid-filled sacks that are licked away by their mothers. Puppies are born both blind and deaf. They gain the use of these senses after a few weeks. Puppies are given milk by their mothers until they are old enough to eat solid food. The milk gives them vital nutrients and helps boost their immune systems.

◗ Reproduction

Female dogs are pregnant for about 60 days. This time is called gestation. They give birth to 1–15 puppies, although the average litter size is usually between 3–6 puppies.

Mothers feed their puppies with milk until they are old enough to eat solid food.

⊃ The role of the mother

Shortly before giving birth, the mother starts to look for a safe place to give birth to her litter. After the puppies are born, the mother instinctively performs a number of tasks to ensure the puppies' survival — such as licking their stomachs to provoke their first reactions. A mother will also naturally reject any deformed or sick puppies.

◗ Wild dog pups

Puppies born in the wild have to compete for survival from the moment they are born. If a puppy is born sick or deformed it will be rejected by its mother. The first year is the most vulnerable time for dogs, because they are often sick and can be singled out by predators as weaker or less experienced than older members of the pack.

Newborn puppies are blind and deaf. They have their first feed shortly after birth. They then spend the majority of their first day sleeping.

At six days old, the skin on the puppy's nose and paws begins to turn darker. At this stage the puppy is still not able to see or hear anything.

Puppies begin to see images after 13 days and their other senses are also developing rapidly. At this stage, the puppy can raise its body off the ground.

By about four weeks old puppies are able to sit and stand like miniature adults. They can also play games and start to socialize with humans.

Children can play an important role in the way that puppies grow up. They can help with many tasks, such as taking the puppy for a walk, brushing, and feeding the dog.

∩ Becoming strong

Playing is a crucial part of a puppy's development. Scientists believe that puppies who do not play enough when they are young will not be able to defend themselves properly when they are older. If you watch a puppy at play, you can see many aspects of adult behavior. Actions such as stalking, chasing, biting, and shaking an object around in its mouth are all part of a dog's hunting instinct.

∩ Puppies and children

Relationships between puppies and children are usually excellent as both tend to communicate with each other through play. Scientists have proven that puppies who have no contact with children are three times more likely to develop illnesses later in their lives.

⊂ Training your puppy

The way a puppy is trained when it is young will affect the way that it behaves in adult life. Most experts think that a puppy should attend an obedience school after it is six months old.

A puppy should be taken to the vet for vaccinations when it is ten weeks old.

⋃ Growing up

The experiences that a puppy has and the environment in which it is placed will all affect the type of dog your puppy becomes when it is an adult. For this reason it is important to establish a firm code of conduct with a puppy from the moment it enters the house.

⋃ First visit to the vet

Newborn puppies receive antibodies from their mother's milk that provide them with a natural immunity to infection. These start to wear off after about eight to ten weeks. At that point, all puppies should be taken to a vet and given injections that will provide them with immunity against various canine diseases.

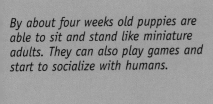

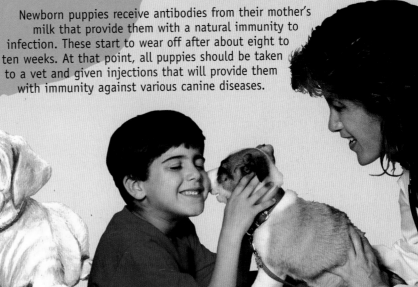

Profile: The Wolf

The relationship between people and wolves is a complicated one. People have always both feared and respected wolves. In many parts of the world, they have tried to eliminate the wolf, while in others they are trying to protect it and the environment in which it lives. Ironically, the wolf is thought to be the direct ancestor of the domestic dog, which is known as "man's best friend."

ᴖ A fearful reputation

People and wolves have been living in close company for centuries and have competed for many of the same sources of food. The wolf's fearsome hunting ability has become legend around the world. Farmers tell stories of wolves that have killed dozens of sheep in a single night. To this day, the wolf appears in many stories and legends as the cunning trickster.

A wolf's howl can travel up to a distance of 5 miles (8 km).

ᴖ Life in the pack

Wolves live in packs, usually made up of members of their extended family and ranging from 5 to 12 in number. Wolves live in a strict hierarchy within this group depending on the number, sex, and age of pack members. The pack is led by a male (called the alpha male) and female (called the alpha female) pair, one of which will be the pack's leader. The job of the leader varies from maintaining the order of the pack to making decisions about the best places to hunt. Half of all wolf packs are led by the alpha female.

ᴖ Wolves in myths

No animal has been enshrined in the myths and legends of Europe more than the Gray wolf. Aesop celebrated the animal's cunning in his fables. Romulus and Remus, the legendary founders of Rome, were kept alive by the milk of a she-wolf, and the Brothers Grimm immortalized the wolf in their story of *Little Red Riding Hood* (above).

ᴖ Communication

Wolves use body language, sound, and smell to communicate with other members of their pack. They also use scent to mark out their territory. The famous howl is used by the wolf as a means of advertising its presence to other members of the pack.

Wolves use a variety of postures and expressions to communicate with one another. Perhaps the most dramatic and aggressive occurs when a wolf pulls back its lips and bares its teeth.

ᴖ Hunting and fighting

Wolves often hunt together as a pack. They are keen observers of behavior and 60% of their kills are made up of the young, the old, or the weak. The pack relentlessly pursues its prey until it becomes too tired to defend itself. An adult wolf can eat up to 20 pounds (9 kg) of meat in a single meal. Fights often occur within the pack, usually as an attempt to reestablish the hierarchy. Challenges for the pack's leadership are most common during the winter months, when food is at its scarcest.

Despite years of persecution and hunting, wolf numbers are rising again in many isolated areas of Europe. The last wolf was killed in Europe's western Alps in 1921, but, due to migration from eastern Europe, six Alpine countries now have a wolf population of more than a thousand. The wolf is now protected species in many countries.

Wolf ID

Scientific name:
  Canidae
Species: 2 species —
  Gray wolf
  (subspecies include:
  Common wolf, Steppe wolf, Tundra wolf,
  Eastern timber wolf, Great Plains wolf,
  Buffalo wolf), and Red wolf
Diffusion: North America, Europe, Asia,
  and Middle East
Status: Buffalo wolf (extinct) and Red wolf
  (probably extinct in the wild)
Height: 40–58 inches (100–150 cm)
Tail length: 13–20 inches (31–51 cm)
Weight: 27–165 pounds (12–75 kg)
Prey: large hoofed animals such as moose,
  elk, deer, sheep, goats, caribou, musk oxen,
  and bison
Gestation: approx. 61–63 days
Life expectancy: 8–16 years
  (20 years in captivity)

Dogs in the Past

Dogs and people have been living together for centuries. Their early associations came through hunting and war. However, the more civilized man has become, the more the role of the dog in his life has changed — it has gone from being a useful animal to have around to its role today, as beloved pet and man's best friend.

This illustration shows a grave in Palestine from about 14,000 years ago in which a boy was buried with a wolf cub.

Recollect that the Almighty, who gave the dog to be the companion of our pleasures and our toils, hath invested him with a nature noble and incapable of deceit.

Sir Walter Scott (1771–1832), author

Dogs were used extensively by the Native Americans for pulling sledges carrying goods — this practice continued even after the arrival of the horse in North America.

◔ Dogs in ancient Egypt

The ancient Egyptians valued dogs and they played an important part in their culture. Dogs were deified and became the symbol of Anubis — the god of the dead who guided human souls into the afterlife (left). Set, the god of evil, was represented with the head of a greyhound. Anyone who mistreated a dog was punished by death.

♁ Early domestication of dogs

People disagree about when, where, and why dogs became pets. Initially, dogs may have been used to help in the hunt or trained to fight in battles. They may also have been used as guard dogs to protect temples and large estates. Alternatively, orphaned wolf cubs living close to villages may have been adopted and brought up by children.

☻ Dogs in the Far East

China does not have a hunting tradition. For this reason, dogs were domesticated at a late stage, mostly through Western visitors who gave dogs as pets to the Emperor. Japan, like China, also had little use for the dog, but it does appear in Japanese religion. Omisto, the god of suicide, was often portrayed with a dog's head.

This Japanese lion dog (left) from the Heian period (794–1185) once stood in the outer hall of a shrine.

♁ From Greece to Rome

Ancient Greeks had dogs from the earliest times, both as pets and for hunting. Dogs were also used in battle after the Greeks defeated Xerxes I of Persia in 480 B.C. However, dog loving as we know it today began in ancient Rome, where various new breeds were developed, including the lap dog.

This Greek bas-relief (above) dating to 510 B.C. shows a dog and a cat fighting.

☻ Dogs in the Middle Ages

The Middle Ages began badly for dogs. After the fall of the Roman Empire, people returned mostly to the countryside and, as a result, dogs returned to a semi-wild state, roaming in savage packs, spreading diseases (mostly rabies), and scavenging for food. Around A.D. 1000, dogs were bred once more for hunting, a popular pastime for the nobility, and sheep dogs were once again revived.

A scene from the Bayeux tapestry (right) shows dogs accompanying men into the Battle of Hastings in 1066.

This mosaic (above), placed at the entrance to a house in Pompeii bore the words "Cave Canem" (Beware of the Dog).

➲ Dogs and prosperity

During the Renaissance, the typical dog was the hunting dog. The first book on various dog types appeared in England in 1590. Owning a dog, once the privilege of the aristocracy, became widespread by the end of the Renaissance, although dogs belonging to people of the lower classes had to have their tendons cut by law to prevent them from hunting.

A portrait of Charles V (1500–1558), king of Spain, and his dog by Titian (c. 1488–1576).

⊜ The different origins of breeds

People have been breeding dogs ever since they first became domesticated. By the end of the 16th century, there were five recognized breeds of dog — greyhounds, terriers, slowhounds, largehounds, and bulldogs. The English were the first to create new breeds and became the world's first major exporter of dogs for pets. The Isle of Dogs in London was set aside by the Tudors as a kennel for the breeding of dogs to cater for the new demand for the animals in Europe. Other countries soon followed, particularly France.

Working Dogs

Ever since dogs first began to live with humans they have been taught to help them with their work. Dogs have joined soldiers on the battlefield throughout history, just as they have helped people protect their livestock. More recently, dogs have been trained to help disabled people, to sniff out illegal drugs, and to rescue people from the water or buried under snow or debris.

A memorial commemorating the deeds of dogs in war at Hartsdale in the United States (left).

⊂ Police dogs

Dogs have been used as an extra arm of the law for years. They can be trained to track criminals, to assist in crowd control, and to help expose traces of illegal substances (such as drugs hidden in suitcases). One dog, a Labrador called Snag, made 188 drug seizures worth $810 million during his career with the U.S. Customs Service.

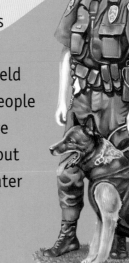

Owning a dog can be hard work if you are not really a dog lover. In that case, there are many robot dogs available, which do not need feeding or walking.

⋂ Assistance dogs

Dogs can be trained to perform a number of tasks for disabled people. They can act as eyes for blind people, as ears for people who are hard of hearing, as well as retrieving objects for people with limited movement. A French lieutenant first came up with the idea for guide dogs during World War I. It takes between four and six months to train a dog to perform such tasks.

⋂ War dogs

Dogs have accompanied soldiers onto the battlefield for centuries. The ancient Egyptians first encountered them when they suffered defeat at the hands of the Hyksos people and their fearsome mastiffs. In recent times, dogs have served in both the First and Second World Wars as well as in the Vietnam War.

⟳ Sheep dogs

Dogs have been used as sheep dogs for as long as people have kept sheep (at least 8,000 years). A sheep dog not only protects the flock from predators, but also moves it to wherever the shepherd wants it to go. A good sheep dog can look after 150 sheep and will be able to cover 20 miles (32 km) a day.

The beautiful Newfoundland (above) has always been used to rescue people from the water.

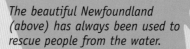

> They [dogs] never talk about themselves but listen to you while you talk about yourself, and keep up the appearance of being interested in the conversation.
>
> *Jerome K. Jerome (1859–1927), author*

↻ Truffle dogs

The truffle is a highly prized fungus that grows beneath the ground. It is often called the "diamond of cooking." Two animals are traditionally used to locate truffles — dogs and pigs. Dogs are the preferred truffle finders because they will locate the delicacy simply for the pleasure of finding it and not because they want to eat it.

Dogs can detect truffles hidden 3 feet, 3 inches (1 m) underground.

Nowadays, dalmations are the preferred fire house mascots.

↻ Avalanche dogs

Dogs have been used to find people buried in the snow for hundreds of years. The most famous breed for performing this task is the Saint Bernard, which takes its name from the hospice in the Alps where it was used to help travelers find the snow-covered path through the mountains. Today, German Shepherds and Golden Retrievers are also used for this task.

A dog's keen sense of smell can be used to find people buried under rubble or snow.

⊃ Rescue dogs

A dog's heightened sense of smell can be put to good use when it comes to finding people buried in rubble, after an earthquake or a landslide, for example. Dogs were used after the World Trade Center attacks on September 11, 2001. They not only provided an invaluable service when it came to finding survivors, but also provided good therapy for the rescuers.

A rescue worker at the World Trade Center disaster with his dog (right).

↻ Fire house dogs

Dogs were originally employed in fire houses to kill rats and to guard the fire station against intruders. They also directed the horses toward the scene of the fire and kept stray dogs away from the equipment as the firefighters were going about their work. There are fewer dogs in fire houses today, although some still serve as mascots.

Hunting

Dogs and people hunting together is an age-old pastime. Dogs can be trained not only to discover prey but also to bring it back to their masters. Some breeds of dog have been created for this very purpose — such as the Golden Retriever. Packs of dog can also be used to accompany men hunting on horseback. This can still be seen today in some parts of the world with fox hunting.

The English aristocracy led the fashion for large-scale hunts with dogs in the 18th century, as depicted here (left) by the English artist George Stubbs.

Hunting and Sporting Dogs

Dogs have found their way into people's homes as pets and have also helped them with their work for centuries. Since ancient Egyptian times, dogs have provided people with entertainment, whether it be by accompanying them on a hunt or through racing. Today there are a number of spectator sports involving dogs.

Musical freestyle

Musical freestyle is a sport that combines dog obedience with dance. The dog and its owner perform a routine set to music and the dog is encouraged to execute a variety of movements and tricks. An international organization for musical freestyle was formed in 1994.

Two breeds of dog are traditionally used in sled dog racing — the Alaskan Malamute and the Siberian Husky.

American freestyler Carolyn Scott and her Golden Retriever Rookie, during a freestyle performance (right).

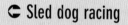

Sled dog racing

The sport of sled dog racing dates back thousands of years. Native American people were dependent on dogs for a number of things, but mostly for transportation. The first official sled dog race took place in Alaska in 1908 and covered a course 408 miles (656 km) long. Enthusiasm for the sport spread rapidly through North America and today it has become globally popular.

↻ Flyball

Flyball is a team relay race with four dogs and their owners in each team. The aim is for the dogs, in turn, to jump over four hurdles, trigger a box which releases a ball, catch the ball and then return to the start-finish line. The next dog is then released to perform the same actions and so on. The winner is the first team to get four balls and four dogs back without a fault.

Agility shows first came about at the Crufts Dog Show in Birmingham in 1977 when something was needed to fill the time between two parts of the competition.

The Kennel Club says that the purpose of dog shows is to promote the general improvement of dogs in every way.

↻ Dog shows

Shows that were devoted solely to dogs started to take place in the 18th century. The most famous dog show in the world is called Crufts and it takes place every year in Birmingham, England. Awards are given to the best dog in each breed (the dogs have to be pedigree dogs) and to the best dog in the show.

↻ Agility shows

Dog agility is a sport where a handler has a set amount of time to guide his or her dog over a series of obstacles. As in show jumping, the dog picks up penalties if it knocks down a bar from one of the hurdles or if it places a foot in a safety zone. The winner is the dog that completes the course with the fewest faults. If two dogs have the same number of faults, the dog that has completed the course in the least amount of time is the winner.

↻ Dog racing

Dog racing has taken place since Roman times, but the sport was truly developed in England in the 16th century, when dogs chasing hares became a popular spectator sport. The hares would be given a head start and two greyhounds would race after it — the sport was called "coursing." The most famous coursing race, the Waterloo Cup, was established in England in 1837.

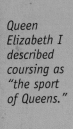

Queen Elizabeth I described coursing as "the sport of Queens."

I've seen a look in dogs' eyes, a quickly vanishing look of amazed contempt, and I am convinced that basically all dogs think humans are nuts.

John Steinbeck (1902–1968), author

Profile: The Fox

The fox has been portrayed by people as a cunning creature that uses its wits to survive and, as a result, has been widely hunted and persecuted. However, the fox should be admired for its alertness, its cleverness, and its keen observation — characteristics that have enabled it to adapt to a vast number of different environments. Fortunately, our attitude to the fox has started to change.

All foxes have very keen senses, and their sense of smell is particularly acute. They are also masters of stealth and camouflage.

A fox's resourcefulness explains why they are so widespread around the world.

⮩ Characteristics

Foxes are small canids with pointed muzzles. They are more widely dispersed than any other mammal (apart from people). They are opportunistic foragers of food whose hunting technique varies from smash and grab to silently stalking their prey. They can be found in a staggering number of different habitats — ranging from the Arctic tundra to the Sahara Desert.

⮩ The fabled fox

The fox is celebrated in many cultures throughout the world. In *Aesop's Fables*, the cunning fox manages to outwit the stork. In medieval tales, Reynard the Fox always manages to get the better of his adversaries and, in the tales of Beatrix Potter, the sly fox manages to get the better of the gullible Jemima Puddleduck. The fox is invariably seen as cunning, deceitful, but, above all, as resourceful.

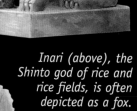

Inari (above), the Shinto god of rice and rice fields, is often depicted as a fox.

⮩ Fox cubs

Foxes normally breed once a year and the vixen (female fox) will carry the cubs in her stomach for a period of 60–63 days (or 51 days for the Fennec fox). The litter can be between one and six cubs. Fox cubs are normally born in burrows or rock crevices. The cubs will remain in the burrow for about two months until they are strong enough to come out into the open.

When they are strong enough, male cubs will be forced to leave the den by the male fox and will travel extensively to establish their own group. Some young males have been known to travel up to 155 miles (250 km) after leaving the den.

Man is the fox's greatest predator, whether it is through fox hunting (left) or for the fur trade.

⮨ Living as a group

Fox society can be a complex one. In some areas, foxes live as a pair, in others, they live in groups of one male and several vixens. The size of the territory a group of foxes inhabits depends almost entirely on the amount of available food and can range in size from 77 square feet (200 sq m) to 25 square miles (65 sq km). Territories may cross over between groups, with the dominant male occupying the best hunting zones.

Fox ID

Scientific name: *Canidae*
Species: 23 in 4 genera: Gray fox; Bat-eared fox; Vulpine fox (Red fox, Swift fox, Arctic fox); and South American zorros
Diffusion: Americas, Europe, Asia, and Africa
Habitat: wide-ranging from Arctic tundra to city centers
Body length: 9.5–39 inches (24–100 cm)
Tail length: 7–14 inches (18–35 cm)
Weight: 2.2–20 pounds (1–9 kg)
Prey: diverse — from small mammals to earthworms, beetles, fish, and fruit
Gestation: from 51 days (Fennec fox) to 60–63 days (Red fox)
Life expectancy: up to 6 years (13 in captivity)

For centuries, the fox has been thought of as a pest because it sometimes attacks poultry or other farm animals for food. For this reason they have been hunted and, in many areas, almost wiped out. Another contributing factor to the reduction of fox numbers has been the fur trade. Attitudes to the fox are changing, however, and their numbers are steadily increasing once again.

↺ Sirius the dog star

According to Greek legend, Canis Major (Big Dog) and Canis Minor (Little Dog) were Orion's two hunting dogs. Sirius, known as the Dog Star, is the diamond in Canis Major's collar and is the brightest star in the heavens. In the past, people thought that Sirius made days hotter. The star was also associated with death and pestilence. Today, it is thought of as a winter star that accompanies Orion on his journey across the heavens.

Osiris (left), the ancient Egyptian god after whom the star was named.

↺ Werewolves

The modern belief in werewolves arose in the 16th century; it involves believing that people can transform themselves into gruesome wolves that can only be killed by a silver object. The word comes from the Greek story of King Lycos who was turned into a wolf by Zeus as punishment for trying to trick him. It led to the word *lycanthrope* — *lycos* meaning "wolf" and *anthropos* meaning "man."

A costume of a werewolf for the theater (right). The actor just had to turn around to show his transformation into a wolf.

Despite attempts to eradicate it, the coyote (below) continues to survive.

↺ Japanese inubako dog

The Japanese inubako dog (right) was thought to help children grow up healthily. It was placed at the side of the bed of a mother who was about to give birth and next to the cradle of the newborn baby. It was believed that the dog had the power to ward off any evil spirits.

Mythical Hounds

The dog has taken on many different guises in myths and legends around the world that celebrate both the animal's good and bad qualities. It could be the ferocious three-headed monster that guards the gates of the Underworld or the loyal, unwavering dog that recognizes its master after an absence of 20 years. The abundance of dog tales shows just how important a role the animal has played in the history of mankind around the world.

↺ Native Americans and dogs

Many of the tribes of the southwestern United States believed that the coyote was the god of tricks. According to the story of the Coyote and the Earthmaker, the coyote is responsible not only for death but for all the bad things that happen in the world. Despite the Earthmaker's orders to kill him, the coyote continually escaped and continues to cause trouble wherever he goes.

A Native American dog dancer (left).

↺ The dog in Christianity

St. Christopher, the patron saint of travelers, was often depicted in paintings with a dog's head. This was because he came from a tribe that the ancient Greeks believed lived outside the civilized world. They called such people "dog-heads."

↺ The Year of the Dog — Chinese New Year

According to Chinese astrology, those born in the Year of the Dog are blessed with the finer traits of human nature. They are loyal, have a deep sense of duty, and can be relied upon to keep a secret. They also tend to be people who support the underdog. Someone born in the Year of the Dog will make a good friend, but they can be somewhat temperamental.

中國
香港

HONG KONG, CHINA

$1³⁰ 狗年生肖 Year of the Dog

☿ Cerberus and Heracles

The 12th and final labor of Heracles was to bring Cerberus, the three-headed dog that guarded the gates of the Underworld, before King Eurytheus. Hades, the god of the Underworld, allowed Heracles to do so as long as he did not use any weapons to subdue the beast. Heracles half killed Cerberus before dragging him to the king, who, terrified, jumped into a large pot to hide.

A bronze dog found at a Celtic healing sanctuary in England (right).

DOG, n. A subsidiary Deity designed to catch the overflow and surplus of the world's worship...His master works for the means wherewith to purchase the idle wag of the Solomonic tail, seasoned with a look of tolerant recognition.

Ambrose Bierce (1842–1914), author

♁ Dogs and Celtic mythology

Since they were scavengers and eaters of carrion, dogs were perceived as messengers of the Otherworld in Celtic mythology. According to the Taliesin poem, Cwn Annwn, Arawn, the King of the Otherworld, owned a pack of dogs that would appear at night to foretell death. Due to this association with the Otherworld, many Celts believed that dogs possessed healing properties. Dogs were also associated with many other deities.

A scene (above) depicting the moment that Heracles brought the three-headed dog Cerberus before King Eurytheus.

♌ Odysseus and his dog

According to Homer's epic story *The Odyssey*, Odysseus's dog, Argos, lay neglected in a pile of dung after his master had left for Troy. When Odysseus returned after his 20-year journey, disguised as a beggar, Argos recognized him at once and wagged his tail. Happy that his master had returned, Argos then died.

A scene (above left) showing the moment that Odysseus returned after his epic 20-year journey and was recognized by his faithful dog, Argos.

The Vikings believed that the monstrous wolf Fenrir (right) would break his chains and devour Odin during the final battle called Ragnarok.

⇒ Fenrir and Ragnarok

Vikings believed in a final battle called Ragnarok, where the gods and their arch-enemies would fight and destroy each other. This final struggle would signal the end of the old world, and a new world would rise out of the disorder. The Vikings believed that at Ragnarok, Fenrir, a monstrous wolf, would break his chains and devour the great god Odin.

Famous Dogs

What exactly makes a dog famous? Some canines have made a name for themselves on the television screen or in the cinema. Others have provided the inspiration for the animator's pen and given generations of children hours of pleasure. Some dogs are true heroes, who have risked their lives to save the lives of hundreds of people. Others have given their lives to allow mankind to pursue scientific advancement, whereas some are simply famous because they are the pets of people in the public eye.

➔ Snoopy

Snoopy was one of the main characters in a cartoon strip called *Peanuts*, created by Charles Schultz (right), that became a cornerstone of American culture. The cartoon beagle became a cultural icon — soldiers in the Vietnam War drew his image on their helmets.

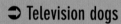

➔ Children's books

Dogs have been the subject of fables since the times of ancient Greece when they appeared in *Aesop's Fables*. In modern times, the most famous literary canine character appears in a book by Dr. Seuss called *Fox in Sox*, the story of a fox who torments poor Mr. Knox with impossible tongue twisters.

◔ Dogs in film

Canine film stars tend to be of the animated kind. Among the most well-known are Walt Disney's *Lady and the Tramp* (1955) and, perhaps most famous of all, *One Hundred and One Dalmatians* (1961), which was made into a feature film in 1996.

Walt Disney's Lady and the Tramp *and* One Hundred and One Dalmatians *were box office hits in the 20th century.*

➔ Television dogs

The first television canine hero was called Rin Tin Tin, a German Shepherd dog that was found in a trench during World War I. He was brought back to Los Angeles and went on to make numerous television series and 22 films. Other canine heroes include Lassie, a collie who first appeared in 1954, The Littlest Hobo, and Mickey Mouse's famous sidekick, Pluto. Dog detective Scooby-Doo (right) and his friends have also made the transition from small screen heroes to cinema stars.

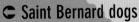

This statue of Balto (below) is the only statue of a dog in New York.

➲ Balto

Balto was a husky who, in 1925, led a dog sled team through blizzards and freezing temperatures to take a vaccine to a community in Alaska. The community was affected by the deadly epidemic diphtheria, which had already killed several children. Balto's epic journey was the subject of an animated film in 1995.

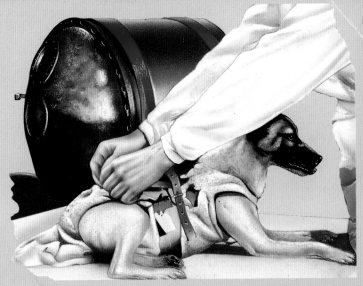

Laika (above) was the first living creature in space.

◑ Saint Bernard dogs

Saint Bernards were used to guide travelers through the snow-covered mountain paths that connected Switzerland and Italy. Their uncanny ability to detect approaching storms and avalanches made them an invaluable rescue dog — they have saved more than 2,500 lives in the past 200 years. Perhaps the most famous Saint Bernard of all was called Barry. Born in 1800, Barry saved the lives of more than 40 people during his 12 years of work.

◑ Dogs in space

In 1957, Russia launched the first satellite to orbit the earth, called *Sputnik I*. A month later, they launched another satellite, this time carrying a dog called Laika (meaning "bark" in Russian), which orbited the earth for eight days. Although Laika did not survive the journey, her trip provided scientists with valuable information for the human space flights which followed.

➲ Famous pets

Queen Elizabeth II of England is famous for her love of corgis. Her affection for the dogs started in 1933 when her father, King George VI, brought two of them home for his daughters. The queen still attends the birth of all the puppies and walks and feeds the dogs herself every day. American presidents have also favored dogs as pets. Lyndon Johnson (36th U.S. president, 1963–1969) was often pictured playing with his dogs. In recent times, George Bush (41st U.S. president, 1989–1993) enjoyed the company of his dog, Millie.

Rin Tin Tin (right) and Lassie (below) were two canines who made it onto the small screen. Rin Tin Tin even had an entry in the telephone directory.

Queen Elizabeth II (above) of England, pictured with one of her much-loved corgis.

◑ Nipper and his master's voice

This image became one of the best-loved trademarks in the United States. Nipper, the star of the picture, became the pet of the Bertrand brothers after the death of his master. One of the brothers, Francis, was a photographer, and he noticed that Nipper would listen carefully to records whenever he played them. It occurred to him that Nipper might be listening for his master's voice and an advertising legend was born.

Domestic Dog Breeds

By the end of the 16th century, there were five recognized breeds of dog: greyhounds, terriers, slowhounds, largehounds, and bulldogs. Today, various Kennel Clubs around the world recognize a different number of breeds. The Kennel Club of the United Kingdom recognizes 192 breeds of dog, which are grouped into six types: hound dogs, gun dogs, terrier dogs, utility dogs, working dogs, and toy dogs. Although cats have taken over as the most popular domestic pet, the dog still has a place in the family home. Here are the ten most popular breeds in the United States.

☊ Dachshund

The Dachshund was originally bred in Germany over 300 years ago. Its name means "badger hound" and it was originally used to flush badgers out of their dens. Dachshunds are fiercely courageous and can be somewhat stubborn. These small dogs, whose legs seem ridiculously out of proportion to the rest of their bodies, are intelligent, active, and fiercely loyal. As such, they make great pets, particularly for those who live in small houses.

⊂ Boxer

Developed in Germany as a guard, working, and companion dog, and introduced to the rest of the world by British and American soldiers after the end of World War II, the Boxer has boundless energy and loves to play. Happy and affectionate, if not a little stubborn, the Boxer is completely devoted to its loved ones and is an ideal companion for children. However, it requires an owner who will love to take it on long walks.

↻ Chihuahua

The smallest breed of dog in the world, even the largest only weighs about 7 pounds (3.5 kg), Chihuahuas have been a popular companion dog for centuries. Loyal by nature, and sometimes even jealous, they are not particularly comfortable with strangers and hate to be left on their own. As they are susceptible to the cold, they should be wrapped up warmly whenever they go outside.

↻ Labrador

The Labrador is a gentle breed with an outgoing nature and a friendly temperament. Originally bred as a gun dog, Labradors can be extremely wild and energetic as pups, but they are courageous and hard-working. Labradors are big dogs with big appetites and need plenty of exercise to keep them from putting on too much weight.

⬅ Golden retriever

Golden Retrievers are a highly intelligent breed that are often used as guide dogs. Their calm, gentle, and loyal nature makes them an ideal pet. Big dogs with large appetites, they need plenty of exercise and shed a lot of hair during the summer months. Bred to retrieve waterfowl in the 19th century, they are natural retrievers who always look for an opportunity to bring things to their owners.

⬅ Beagle

A smaller version of the foxhound, Beagles are great trackers and great hunting dogs. However, they should never be let off their leads, or they will pick up the nearest scent and follow it to wherever it takes them. As a result of their pack hunting instincts, Beagles need canine companionship. Their long ears need to be checked regularly and they need to be brushed once or twice a week.

⬇ Rottweiler

Rottweilers are a massive, powerful breed that make excellent guard dogs. They need an owner with a firm hand who will gain the dog's respect. If this is the case, they should never be vicious and will make an intelligent and devoted pet. Rottweilers were used as an auxiliary to the German army during World War I.

↻ Poodle

The national dog of France, poodles come in three varieties: standard, miniature, and toy. They are intelligent, loving, active, and playful and make excellent pets. Their coat has to be clipped, and with the right style and cut they can be made to look extremely elegant.

↻ German shepherd

German shepherds are curious, intelligent, faithful, and obedient dogs that love to be involved in all family activities. They are big dogs that should not be kept in an apartment or a small house unless they are given plenty of exercise. They are also huge eaters and will eat their own weight in food every two and a half days.

↻ Yorkshire terrier

Originally bred in the 19th century to control rats in the Yorkshire mines, these dogs became very popular with aristocratic ladies. They have dynamic personalities, great courage, and are lovable, clever, and affectionate. Although they are independent by nature, they will make a good companion.

Caring for Dogs

Buying a dog as a pet can be one of the most rewarding decisions anyone can ever make, but it is not a decision to be taken lightly. A pet dog requires an owner's constant attention and will also cost money. When someone is thinking of buying a dog, he or she should make sure to be in a position to give it the best life it can possibly have. That involves time to take it for walks, brushing and bathing it, as well as having enough money for food and veterinary costs.

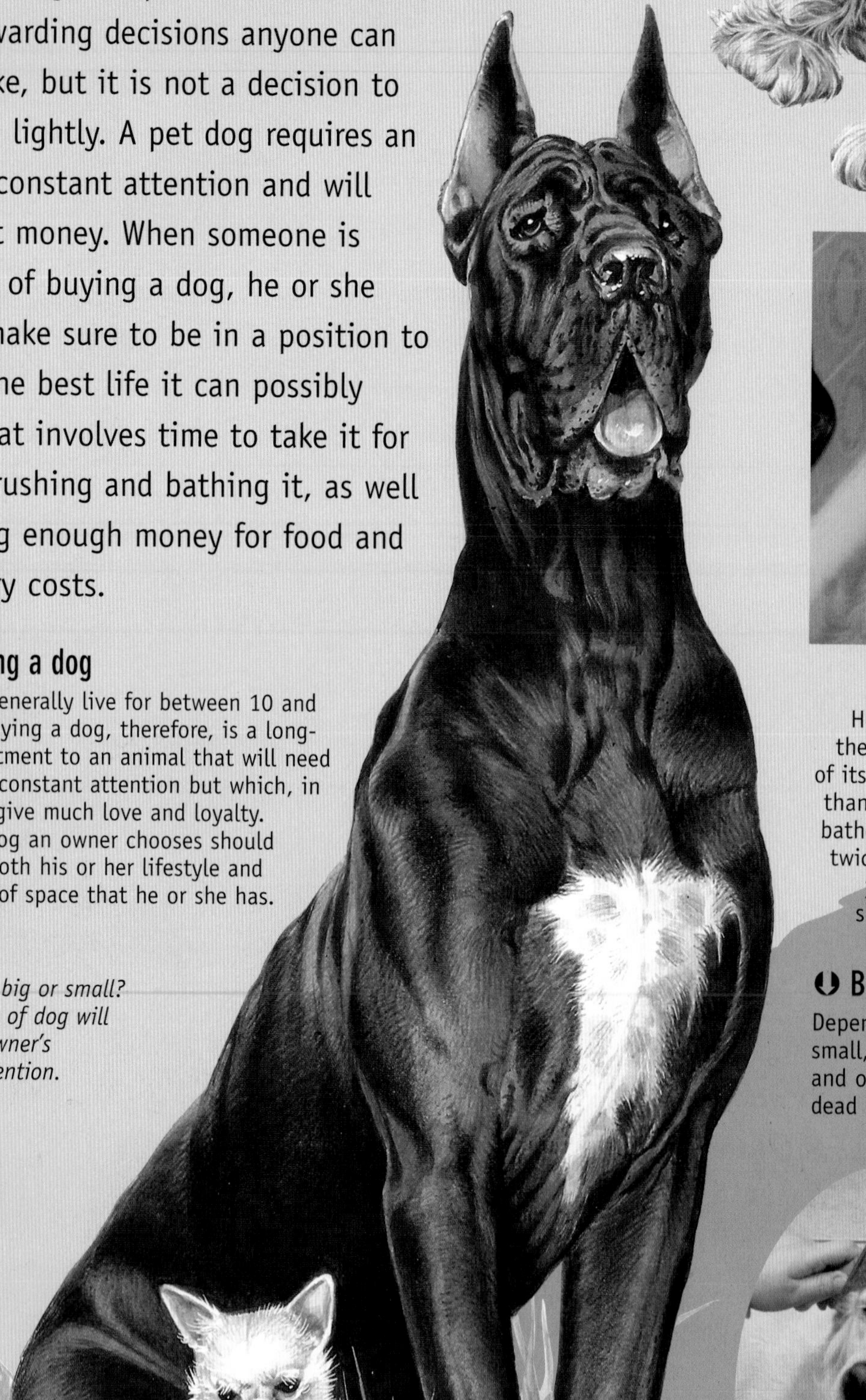

➲ Choosing a dog

A dog will generally live for between 10 and 15 years. Buying a dog, therefore, is a long-term commitment to an animal that will need the owner's constant attention but which, in return, will give much love and loyalty. Whichever dog an owner chooses should fit in with both his or her lifestyle and the amount of space that he or she has.

Should it be big or small? Either choice of dog will require an owner's constant attention.

Bathing

How often a dog needs to be bathed depends on the lifestyle that it leads. A dog that spends most of its time indoors will need to be bathed less often than one that is outside. Certain breeds need to be bathed every couple of months, others only once or twice a year. The best thing to do is to check with your vet when you buy your dog. A special dog shampoo should always be used to bathe a dog.

Brushing

Depending on the breed, all dogs, whether large or small, will need to be brushed between once a week and once a day. Brushing will remove dirt, soil, and dead hair from the dog's coat. All dogs need to be brushed more regularly in spring and in autumn, because that is when they tend to shed more hair. Large dogs with long coats will shed huge amounts of hair. The correct equipment should always be used for brushing.

Golden Retrievers need to be brushed regularly.

✪ The rights of dogs

If someone buys a dog as a pet, it is his or her responsibility to ensure that it leads a full and happy life—if you do so, you will get a huge amount of love and pleasure in return. Too many people buy dogs without realizing how much of their time they require and, each year, an alarming number of pets are abandoned by owners who no longer have the patience to look after them.

As the British Royal Society for the Prevention of Cruelty to Animals (RSPCA) says, a dog is for life and not just for Christmas.

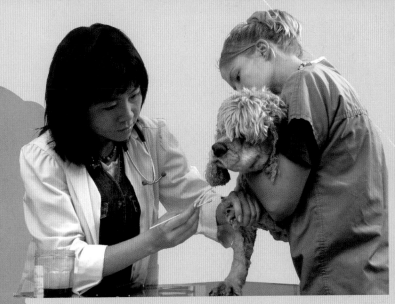

♀ Medical care

Dogs, like any other animal, need to be looked after if they are to live a happy life and if the owner is to get the best out of them. All puppies should be taken to the vet so that they can be vaccinated against canine diseases. If an owner is concerned about his or her dog, or needs advice, a vet is always the best person to ask.

➲ Sleeping

A dog must always have its own place to sleep, be it indoors or outdoors, and this place should be convenient for both the owner and the dog. Indoors, dogs should be kept in an uncarpeted room and given a box with cushions or pillows. Always place the box next to a wall, but not near a radiator. Outdoors, a kennel should be well ventilated and dry. The doorway of the kennel should be one and a half times the size of the dog.

➲ Feeding dogs

Dogs are primarily meat-eaters, but they require a variety of foods for optimum health. A balance of foods ranging from meat and vegetables to grains is a good idea. The amount of food that a dog requires to live a long and healthy life depends very much on its size. A fully grown dog should only have two meals a day and food should never be served at more than 100˚F (38˚C).

Vets will always offer the best advice on a dog's diet. In general, commercial dog foods will provide your dog with a balanced diet.

Index